The INSIDE GUIDE

EARTH'S EXTREME PLACES

Deserts

By Anna Collins

Cavendish Square

Published in 2026 by Cavendish Square Publishing, LLC
2544 Clinton Street Buffalo, NY 14224

First Edition

Website: cavendishsq.com

Library of Congress Cataloging-in-Publication Data

Names: Collins, Anna, author.
Title: Deserts / Anna Collins.
Description: Buffalo, NY : Cavendish Square Publishing, 2025. | Series: The inside guide. Earth's extreme places | Includes index.
Identifiers: LCCN 2024050820 | ISBN 9781502674388 (library binding) | ISBN 9781502674371 (paperback) | ISBN 9781502674395 (ebook)
Subjects: LCSH: Deserts–Juvenile literature. | Desert ecology–Juvenile literature.
Classification: LCC GB612 .C66 2025 | DDC 551.41/5–dc23/eng/20241204
LC record available at https://lccn.loc.gov/2024050820

Editor: Jennifer Lombardo
Copyeditor: Nicole Horning
Designer: Deanna Lepovich

The photographs in this book are used by permission and through the courtesy of: Cover Sabbir Digital/Shutterstock.com; p. 4 muratart/Shutterstock.com; p. 6 Brent Coulter/Shutterstock.com; p. 7 Mike Buchheit/Shutterstock.com; p. 8 kingma photos/Shutterstock.com; p. 9 Sutthi Chuvichit/Shutterstock.com; p. 10 iryna shorokhova/Shutterstock.com; p. 12 udaix/Shutterstock.com; p. 13 (top) Yuliya Zalnova/Shutterstock.com; p. 13 (bottom) Matthew James Ferguson/Shutterstock.com; p. 16 NcikName/Shutterstock.com; p. 18 Ta2ng Ko/Shutterstock.com; p. 20 Griffin Gillespie/Shutterstock.com; p. 21 (main) MyImages - Micha/Shutterstock.com; p. 21 (inset) Bethrus/Shutterstock.com; p. 22 Ion Mes/Shutterstock.com; p. 24 John Theodor/Shutterstock.com; p. 25 EWY Media/Shutterstock.com; p. 27 (main) DavidRebata/Shutterstock.com; p. 27 (inset) Vladimir Zhoga/Shutterstock.com; p. 28 (top) John D Sirlin/Shutterstock.com; p. 28 (bottom) Paulo Miguel Costa/Shutterstock.com; p. 29 (top) Kamran Ali/Shutterstock.com; p. 29 (bottom) Michele Vacchiano/Shutterstock.com.

CPSIA compliance information: Batch #CSCSQ26: For further information contact Cavendish Square Publishing LLC at 1-877-980-4450.

Printed in the United States of America

CONTENTS

The Sahara has been featured in many TV shows and movies, which makes it easy to picture when someone thinks of a desert.

Chapter One

DEFINING DESERTS

For many people, the word "desert" immediately calls to mind huge sand dunes, camel **caravans**, intense heat, and no plants. In other words, their mental picture looks like parts of the Sahara in North Africa. However, there are many deserts around the world that look nothing like the Sahara. For example, the Chihuahuan Desert in the southwestern United States and Mexico is full of plant life. Chile's Atacama Desert has sandy soil and huge, rocky **outcroppings**. Some deserts, such as Antarctica, are even cold and snowy!

What Is a Desert?

Scientists define a desert as an area of land that receives fewer than 10 inches (25 cm) of precipitation per year. Any plants and animals that live in these places have adapted to the dryness. Although many deserts are hot, some can get very cold.

Fast Fact

Antarctica is the largest desert on the planet. It covers about 5.5 million square miles (14.2 million sq km).

In addition to low rainfall, deserts often have very low **humidity**. Because humidity

The Sonoran Desert in the southwestern United States and northwestern Mexico has the most vegetation of any desert in the world.

traps heat from the sun in the air, having low humidity means the temperature in some deserts can change **dramatically** when the sun is no longer out. For example, the difference between day and night in the Sahara can be as much as 36 degrees Fahrenheit (20 degrees Celsius).

Wind and Rain

A desert's low humidity is part of what keeps it dry. Clouds rarely form in the dry air. When they do, light rainfall **evaporates** in the air before it ever reaches the ground. Some deserts get all of their yearly rain from a single large storm, which is known as a cloudburst.

Heavy rainfall can create temporary shallow lakes, called basins, as water gathers in dips in the ground called depressions. Eventually,

EXPANDING DESERTS

Our planet is always changing, and deserts are no exception. Some of the world's most iconic deserts didn't exist a few thousand years ago. For example, the Sahara was full of rivers, forests, and animals about 10,000 years ago. It experienced natural dry periods, but it always recovered—until about 8,000 years ago. At that point, the balance tipped, and the Sahara gradually became the sandy stretch we know today.

The change from grassy fields to dry desert is called desertification. Although this can happen naturally, human activities are making desertification happen much faster than is naturally possible. Since 1920, the Sahara has expanded by about 10 percent, and it is still expanding today.

the water evaporates from the basin, leaving behind playas—dry lake beds covered in salt. Playas are also known as salt flats.

Some deserts experience winds that are up to 60 miles

Cloudbursts can cause quick-moving flash floods, which are very dangerous to hikers.

Camels have thick eyelashes that protect their eyes from windblown sand.

Fast Fact

Cold deserts get most of their precipitation from snow rather than rain.

(100 km) per hour. In a forest or city, plants or buildings block high winds so they lose some of their power. In the desert, however, there is almost nothing around to block the wind. Windstorms create sandstorms, whipping up the sand and swirling it around. This can reduce visibility the same way a blizzard does. In the Sahara, the winds are sometimes so powerful that they carry dust across the ocean, changing the color of the sky as far east as the United States.

Hot and Cold

Some deserts—including the Sahara, Namib, and Kalahari—are classified as hot deserts. These are generally warm or hot at all times of the day and year. They have very few plants, and they are located in the subtropical parts of the world.

Cold deserts can also be hot, especially during summer days. However, they get very cold at night and throughout the winter. Cold deserts are generally found in more **temperate** areas than hot deserts. Some examples are the Gobi Desert in parts of Mongolia and China, the Great Basin Desert in the southwestern United States, and parts of Patagonia in Chile.

Fast Fact

Antarctica and the Arctic region are a kind of cold desert known as a polar desert.

Snowy winters are common in the Gobi Desert.

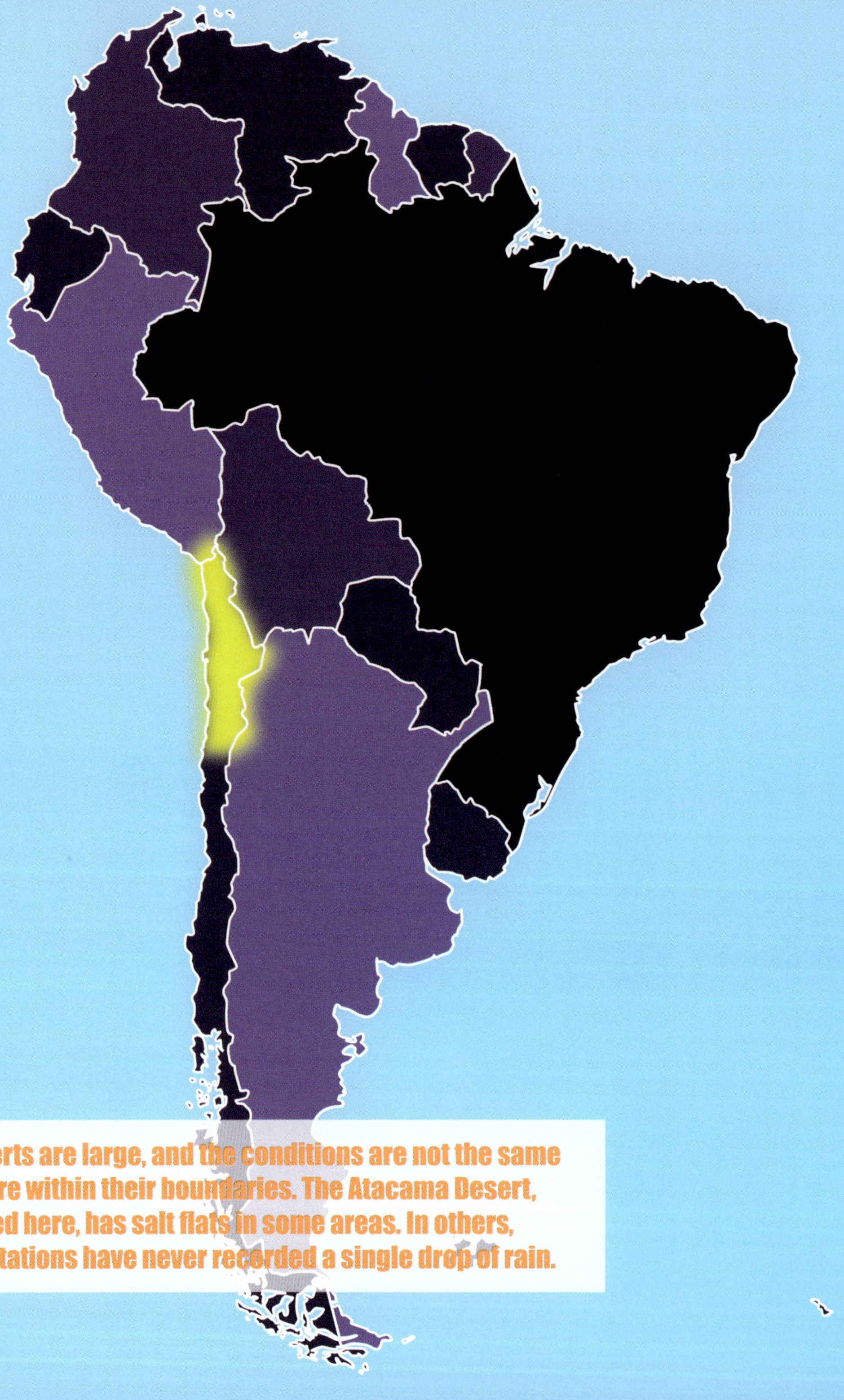

Most deserts are large, and the conditions are not the same everywhere within their boundaries. The Atacama Desert, highlighted here, has salt flats in some areas. In others, weather stations have never recorded a single drop of rain.

Chapter Two

TYPES OF DESERTS

Deserts cover about one-fifth of Earth. Some deserts—including the Sahara, the Arabian, and the Gobi—are well-known around the world. These deserts are sandy and have very little plant life. However, such deserts account for only about 10 percent of the world's total deserts. There are five different types of deserts in the world: subtropical, coastal, rain shadow, interior, and polar. These names **describe** the cause of each desert's dryness.

Subtropical

Subtropical deserts are found near the Tropic of Cancer and the Tropic of Capricorn. They are caused by the circulation, or movement, of large amounts of air called air masses. Near the equator, hot, moist air masses rise in the atmosphere. They cool down as they rise, which causes heavy rainfall and makes the air mass cooler and drier.

As the air mass moves away from the equator, it falls and warms up again, but it does not refill with water. The result is a hot, dry air mass over deserts such as the Sahara and the Kalahari, both of which are in Africa.

Fast Fact

The Sahara covers an area that is just about the same size as the continental United States.

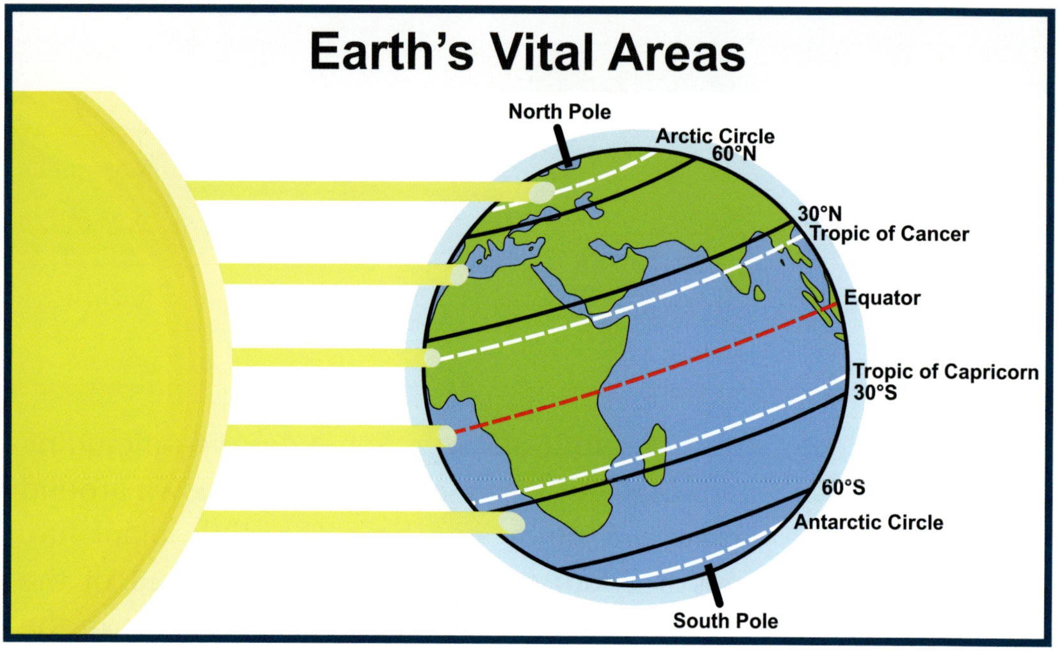

This picture shows the Earth's tropics and the angle at which sunlight strikes them.

Coastal and Interior

As their name suggests, coastal deserts form along coastlines. Ocean currents send cold air toward the shore, where it forms fog. Because fog is made up of water droplets, it creates humidity in the air. However, fog is not a form of precipitation. Coastal deserts rarely have the right air conditions for rainfall. This means that although the air is humid, the ground remains dry.

In contrast, interior deserts form so far inland, or away from the ocean, that the air has lost all its moisture by the time it reaches these

Fast Fact

The only two coastal deserts in the world are the Atacama Desert in Chile and the Namib Desert in southern Africa.

The Atacama Desert is the driest place on Earth.

areas. The Patagonian Desert and Gobi Desert are two examples of interior deserts.

Rain Shadow

Rain shadow deserts form on the leeward side of some mountain ranges. When a humid air mass reaches a mountain range, it rises above the peaks and forms clouds, which deposit precipitation. This is why

Death Valley is a rain shadow desert on the leeward side of the Sierra Nevada mountain range. It is the driest place in North America.

Fast Fact

"Leeward" describes the side of a mountain that faces away from the wind.

MYSTERIOUS MOVING ROCKS

For many years, the "sailing stones" of Racetrack Playa in Death Valley have mystified visitors and researchers. These rocks, which range in size from small pebbles to large boulders, move on their own for no apparent reason. No one has ever seen them move, but they leave tracks behind to show where they have been. Some go only a short distance, but others travel up to 1,500 feet (0.5 km) from their original location.

In 2014, researchers used time-lapse photography to try to solve the mystery. Although they can't be completely sure, the experiment suggests that the rocks move when wind, rain, and ice are all present in the right amounts. The rain creates a sheet of ice in cold temperatures, and as the ice sheet melts and breaks up, wind pushes the pieces of ice along with the rocks on the ground.

mountain peaks often have snow on them. As the air moves past the peak and toward the leeward side, it sinks and warms up.

Because this warming air has already dropped its moisture on the mountain, it is very dry by the time it gets to the leeward side. These conditions make it difficult for clouds to form, creating a desert in the shadow of the mountain range.

Polar

There are only two polar deserts in the world. They are in the Arctic and Antarctica. Large parts of these two polar regions are covered in ice,

Desert Type	Average Temperature	Common Plants	Common Animals
subtropical	summer: 84-95°F (29-35°C) winter: 59-77°F (15-25°C)	evergreen trees, palm trees, bushes	camels, lizards, fennec foxes, tortoises
inland	day: up to 100°F (38°C) night: down to 25°F (-4°C)	cacti, wildflowers, yucca	scorpions, spiders, mice
coastal	annual: less than 68°F (20°C)	grasses, bushes	coyotes, badgers, owls, eagles
rain shadow	summer: up to 100°F (38°C) winter: down to -36°F (-38°C)	flowering plants, prickly bushes	flying insects, rabbits, bighorn sheep
polar	annual: less than 50°F (10°C)	mosses, lichens, grasses	polar bears, Arctic foxes, walruses, penguins

Shown here are some examples of the features of different kinds of deserts.

with little to no snow. Although ice is a form of water, the ice sheets in polar deserts never melt under normal conditions, so they never release moisture into the ground.

Fennec foxes have many adaptations to help them survive in the desert. Their large ears cool them down by releasing heat as blood flows through them. At night, when temperatures drop, their thick coat keeps them warm. They also have special kidneys that help them retain, or keep, as much water in their body as possible.

NATURAL ADAPTATIONS

Chapter Three

It might seem as though nothing could possibly survive in a desert. However, plants and animals have adapted to the harsh climates of the world's deserts. Their adaptations are passed down through their **genes**.

Plants

There are thousands of different kinds of plants that live in deserts all over the world. The one thing they have in common is that they all have adaptations that allow them to survive on very little water. One common adaptation is small leaves or no leaves at all. Plants release water from their leaves into the air through a process called transpiration. The smaller the leaf, the less surface area there is for transpiration to take place—which means the plant retains more water. Desert plants' leaves often have a waxy coating on them that gives them extra help retaining water.

Another adaptation some plants have is in their roots. Some plants have a wide, shallow system of roots. This lets the roots spread out far from the plant to get any water

Fast Fact

Animals that have adapted to life in the desert are called xerocoles.

Instead of leaves, most cacti have prickly spines. The spines have the added benefit of stopping some animals from eating the cacti.

Fast Fact

In most plants, photosynthesis takes place in the leaves. In plants without leaves, this process takes place in the plant's stem.

that may be in the surrounding soil. Others have a thick root called a taproot that extends much deeper than typical roots. A taproot can get to the water that sinks into the soil and collects underground.

CLIMATE CHANGE

Human activities are making the planet heat up much faster than it would normally. This is causing serious changes to Earth's climate in many different ways. In deserts that are already hot, the increased heat can cause a drought, or a very long period without rain. Plants and animals in the desert depend on the little rain that falls. Without it, they have a much harder time surviving, even with their adaptations.

Climate change is also melting the ice sheets in polar deserts. In fact, these areas are warming faster than any other place on Earth. This is a problem for animals such as polar bears and penguins. In many places, the ice sheets cover water. Without them, these animals have nowhere to stand, and they aren't adapted to be in the water all the time.

Fast Fact

Camels don't store water in their humps. Instead, they store fat. This helps them survive when there is little food to be found. Camels have other ways of retaining water, such as not peeing very often.

Animals

Like plants, some desert animals store water. For example, desert tortoises can store water in their bladder. Other animals, such as the kangaroo rat, get all the water they need from their food.

Animals also have adaptations to help them stay cool or keep warm based on the kind of desert they live in. For example, the Costa's hummingbird survives cold nights in the Sonoran and Mojave deserts by entering a state of torpor. This is similar to hibernation but doesn't last as long. When the temperature drops, so does the bird's heart rate—from between 500 and 900 beats per minute to just 50. This helps the hummingbird save its energy for maintaining its body temperature. Snow leopards, which live in the cold Gobi Desert, have thick coats to keep them warm.

One of the kangaroo rat's main food sources is seeds from various plants. Its body is designed to store the moisture from those seeds.

There are many more animals in the Arctic polar desert than there are in Antarctica. Some of these include polar bears, Arctic foxes, and walruses. Arctic foxes have thick fur all over, even on their feet. Walruses have a thick layer of fat, called blubber, under their skin that keeps them warm.

Polar bears have black skin and two layers of fur (*below*). The layer next to their skin is short and fuzzy. The top layer is made up of longer, hollow, transparent hairs. Being hollow helps the hairs trap heat better. The transparent fur looks white to our eyes, much like a bundle of transparent straws (*right*) appears white.

No country owns Antarctica, but several countries have scientific bases there. Shown here is the Chilean base.

Chapter Four

EXTREME LIVING

People have also adapted to desert living. While animals have naturally changed over millions of years, humans have adapted more quickly with special clothes, houses, and tools. There are people living in both hot and cold deserts all over the world. Their ways of life depend a lot on the type of desert they live in. People can also visit and explore deserts—if they have the proper gear!

Hot Deserts

People have found many different ways to live in hot deserts. Clothing choice is one very important part of desert life. In the Saharan and Arabian deserts, people often wear clothes that cover most of their bodies but are light and loose. This protects them from sunburn while also allowing airflow so they don't get too hot. Because light colors reflect light and dark colors absorb it, their clothing is generally white or light gray.

> **Fast Fact**
>
> About 1 billion people live in deserts around the world.

Housing is also important in deserts. In the Sonoran Desert, the Anasazi people **traditionally** made

A kaffiyeh is a square of fabric that is worn on top of the head to protect it from the sun. A kaffiyeh can be pulled up around the nose and mouth to protect the wearer during sandstorms.

The Anasazi sometimes built their homes in caves on the sides of cliffs. People can visit one of the most famous Anasazi villages at Mesa Verde National Park in Colorado.

their homes out of mud bricks, and the walls were thick to keep them **insulated**. This meant that the temperature inside the buildings stayed fairly comfortable, day or night, even as the temperature outside changed dramatically. They had windows to let in air and light, but these were small to limit the amount of sand and dust that could get in.

Cold Deserts

In some deserts, the lack of plants and water means people have had to become nomadic. This means they travel from place to place hunting

OASES

Although deserts are generally dry, there are some places where fresh water from underground comes to the surface to form small bodies of water. These are called oases. The water in an oasis allows for the growth of plants that are not typically found in deserts, such as figs, olives, and oranges.

Some oases are created naturally. Others are created by desert nomads as places to get fresh water for themselves and their animals, as well as to plant certain crops they can harvest over time. Because food and water are so hard to find in the desert, nomads are careful to maintain oases. They have become important stops on trade routes, and people sometimes fight for control over a specific oasis.

Fast Fact

Only scientists live in Antarctica year-round. However, several cruise lines offer Antarctic vacations.

and gathering, rather than settling down and farming. Nomads need houses that are easy to move, but still protect them from the weather. In places such as the Gobi Desert, many people live in tents. These tents have walls that are thick enough to keep people warm in cold weather but have openings to allow breezes to flow through on hot days.

About 4 million people live in the Arctic, and about 10 percent of them belong to various Indigenous, or Native, groups. One well-known group is the Sami. The Sami people have traditionally made their living herding reindeer. They eat the meat, use the skin

In Mongolia, nomads live in round tents called yurts or gers. However, not all Mongolians are nomads.

to make warm clothing, and use the antlers and bones to make tools and decorations.

Today, many Sami people live in homes similar to the ones found in most other parts of the world. However, in the past, they made homes out of wood and covered them with dirt and grass. This insulated the home to keep it warm, even when the temperature outside dipped below freezing. No matter what the conditions are like, desert dwellers have learned how to live comfortably!

Fast Fact

Other Indigenous Arctic groups include the Inuit, Nenet, Chukchi, and Yupik. Overall, there are more than 40 different Indigenous Arctic groups.

THINK ABOUT IT!

1. Why do you think hot, sandy deserts are what most people think of when they hear the word "desert"?

2. Which type of desert are you most familiar with?

3. What other adaptations can you think of that
would help a plant or animal survive in a desert?

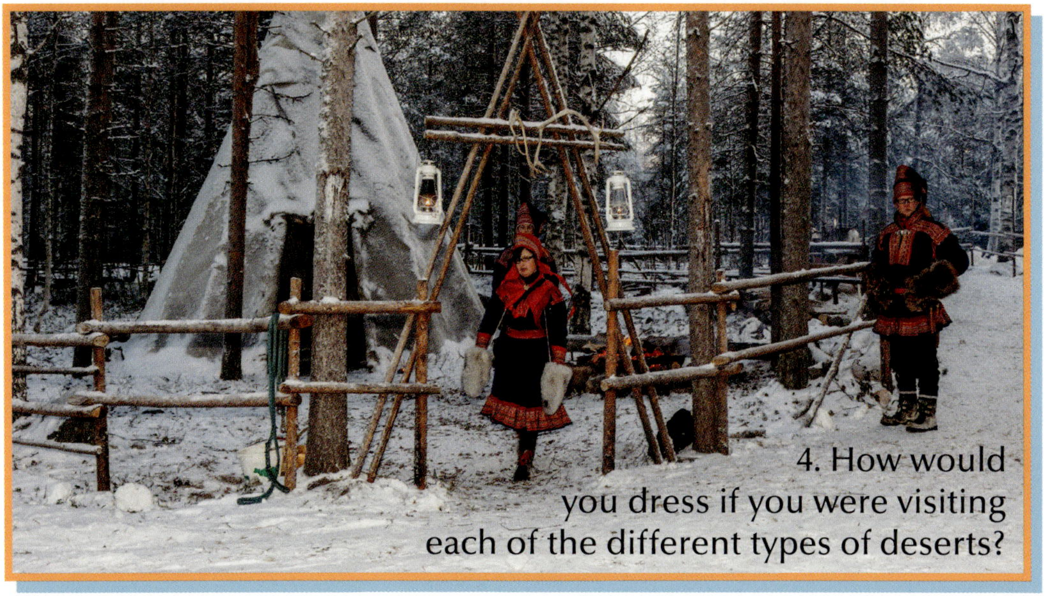
4. How would
you dress if you were visiting
each of the different types of deserts?

GLOSSARY

caravan: A group of pack animals or vehicles traveling one behind the other.

describe: To give an account of something in words.

dramatically: To a very great and often surprising degree.

evaporate: To pass from a liquid state to a gas state.

gene: One of many parts of a cell that control the appearance, growth, and other qualities of a living thing.

humidity: Moisture in the air.

insulate: To use a material to keep the temperature of something constant while the temperature of the air around it changes.

outcropping: The part of a rock formation that appears aboveground.

photosynthesis: The process by which plants turn light, carbon dioxide, and water into energy.

temperate: Having a climate that is neither very hot nor very cold.

traditionally: According to what has been done for a long time.

transparent: See-through.

FIND OUT MORE

Books

Gagne, Tammy. *Let's Explore Deserts*. Mankato, MN: Wonder Books, 2023.

Green, Sara. *Deserts*. Minneapolis, MN: Bellwether Media, 2022.

Roumanis, Alexis. *Deserts*. New York, NY: Lightbox Learning, 2024.

Websites

Ducksters: Desert Biome
www.ducksters.com/science/ecosystems/desert_biome.php
Read more cool desert facts.

Kiddle: Desert Facts for Kids
Kids.kiddle.co/Desert
Learn more about deserts, and take a look at some amazing pictures.

SciShow Kids: Amazing Ways to Live in the Desert!
www.youtube.com/watch?v=gaZKEc59g1w
This video explores some of the ways plants, animals, and people have adapted to desert life.

Publisher's note to educators and parents: Our editors have carefully reviewed these websites to ensure that they are suitable for students. Many websites change frequently, however, and we cannot guarantee that a site's future contents will continue to meet our high standards of quality and educational value. Be advised that students should be closely supervised whenever they access the internet.

INDEX